Collins

Maths
in 5 minutes

Quick practice activities

Chief editor: Zhou Jieying
Consultant: Fan Lianghuo

CONTENTS

HOW TO USE THIS BOOK

The best way to help your child to build their confidence in maths and improve their number skills is to give them lots and lots of practice in the key facts and skills.

Written by maths experts, this series will help your child to become fluent in number facts, and help them to recall them quickly – both are essential for succeeding in maths.

This book provides ready-to-practise questions that comprehensively cover the number curriculum for Year 4. It contains 40 topic-based tests, each 5 minutes long, to help your child build up their mathematical fluency day-by-day.

Each test is divided into three Steps:

- **Step 1: Warm-up (1 minute)**
  This exercise helps your child to revise maths they should already know and gives them preparation for Step 2.

- **Step 2: Rapid calculation ($2\frac{1}{2}$ minutes)**
  This exercise is a set of questions focused on the topic area being tested.

- **Step 3: Challenge ($1\frac{1}{2}$ minutes)**
  This is a more testing exercise designed to stretch your child's mental abilities.

Some of the tests also include:

- a Tip to help your child answer questions of a particular type.
- a Mind Gym puzzle – this is a further test of mental agility and is not included in the 5-minute time allocation.

Your child should attempt to answer as many questions as possible in the time allowed at each Step. Answers are provided at the back of the book.

To help to measure progress, each test includes boxes for recording the date of the test, the total score obtained and the total time taken.

ACKNOWLEDGEMENTS

The authors and publisher are grateful to the copyright holders for permission to use quoted materials and images.

All images are © HarperCollins*Publishers* Ltd and © Shutterstock.com

Every effort has been made to trace copyright holders and obtain their permission for the use of copyright material. The authors and publisher will gladly receive information enabling them to rectify any error or omission in subsequent editions. All facts are correct at time of going to press.

Published by Collins in association with East China Normal University Press

Collins
An imprint of HarperCollins*Publishers*
1 London Bridge Street
London SE1 9GF

HarperCollins*Publishers*
Macken House, 39/40 Mayor Street Upper,
Dublin 1, D01 C9W8, Ireland

ISBN: 978-0-00-831111-7

First published 2019
This edition published 2020
Previously published as Letts.

10 9 8 7 6 5

©HarperCollins*Publishers* Ltd. 2020,
©East China Normal University Press Ltd.,
©Zhou Jieying

British Library Cataloguing in Publication Data.

A CIP record of this book is available from the British Library.

Publisher: Fiona McGlade
Consultant: Fan Lianghuo
Authors: Zhou Jieying, Chen Weihua and Xu Jing
Editors: Ni Ming and Xu Huiping
Contributor: Paul Hodge
Project Management and Editorial: Richard Toms, Lauren Murray and Marie Taylor
Cover Design: Sarah Duxbury
Inside Concept Design: Paul Oates and Ian Wrigley
Layout: Jouve India Private Limited

Printed and bound in the UK

MIX
Paper | Supporting responsible forestry
FSC™ C007454

This book is produced from independently certified FSC™ paper to ensure responsible forest management.

For more information visit:
www.harpercollins.co.uk/green

1 1000 more or less

Date: _____

Day of Week: _____

STEP 1 (1 min) Warm-up

Start the timer

Answer these.

23 + 10 = [ ]   428 – 10 = [ ]   504 + 10 = [ ]   667 – 10 = [ ]

346 + 100 = [ ]   473 – 100 = [ ]   88 + 100 = [ ]   299 – 100 = [ ]

970 + 100 = [ ]   901 – 100 = [ ]

STEP 2 (2.5 min) Rapid calculation

Start the timer

Answer these.

30 + 1000 = [ ]   1100 – 1000 = [ ]   90 + 1000 = [ ]

2300 – 1000 = [ ]   1400 + 1000 = [ ]   4700 – 3700 = [ ]

4650 + 1000 = [ ]   8460 – 1000 = [ ]   3390 – 1000 = [ ]

9309 – 1000 = [ ]   5781 – 1000 = [ ]   5077 + 1000 = [ ]

4356 + 1000 = [ ]   1400 + 1000 = [ ]   9720 + 1000 = [ ]

STEP 3 (1.5 min) Challenge

Start the timer

Fill in the missing numbers.

740 + [ ] = 840   6352 + [ ] = 7352

5871 – [ ] = 5771   4867 – [ ] = 4857

3499 + [ ] = 4499   3038 + [ ] = 3138

9548 – [ ] = 8548   4895 – [ ] = 4885

Time spent: _____ min _____ sec. Total: _____ out of 33

Date: _____

Day of Week: _____

STEP 1 (1 min) Warm-up

Start the timer

Write each of these numbers in words on a separate sheet of paper.

7326	3821	8230	9526
9662	5600	3702	8714

STEP 2 (2.5 min) Rapid calculation

Start the timer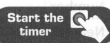

Fill in the missing numbers. The first one has been done for you.

6428 = **6000** + **400** + **20** + **8**   7788 = ☐ + ☐ + ☐ + ☐

1560 = ☐ + ☐ + ☐ + ☐   2800 = ☐ + ☐ + ☐ + ☐

3395 = ☐ + ☐ + ☐ + ☐   9065 = ☐ + ☐ + ☐ + ☐

4586 = ☐ + ☐ + ☐ + ☐   5080 = ☐ + ☐ + ☐ + ☐

3499 = ☐ + ☐ + ☐ + ☐   7086 = ☐ + ☐ + ☐ + ☐

8550 = ☐ + ☐ + ☐ + ☐   6034 = ☐ + ☐ + ☐ + ☐

STEP 3 (1.5 min) Challenge

Start the timer

Write these numbers in digits.

2 thousands and 1 hundred	
9 thousands, 8 hundreds and 5 tens	
4 thousands, 6 hundreds, 2 tens and 6 ones	
7 thousands and 3 ones	
5 thousands and 3 tens	
4 thousands, 1 hundred and 6 ones	
6 thousands, 8 hundreds and 2 tens	
9 thousands and 4 hundreds	

Time spent: _____ min _____ sec. Total: _____ out of 27

Date: _____

Day of Week: _____

1. Find the nearest whole tens.

[ ] ← 3901 → [ ]   [ ] ← 5831 → [ ]   [ ] ← 4405 → [ ]

2. Find the nearest whole hundreds.

[ ] ← 3290 → [ ]   [ ] ← 4582 → [ ]   [ ] ← 5349 → [ ]

3. Find the nearest whole thousands.

[ ] ← 9804 → [ ]   [ ] ← 8313 → [ ]   [ ] ← 2500 → [ ]

Fill in the missing numbers.

Number	4972	3021	2389	8951	9314
Nearest whole tens					

Number	3401	4718	7382	6478	5639
Nearest whole hundreds					

Number	4890	3056	2795	4200	5841
Nearest whole thousands					

Find the pattern and fill in the missing numbers for each sequence.

2150, 2200, 2250, [ ], [ ]   7685, 7585, [ ], [ ], 7285

9753, 7753, [ ], [ ], 1753   3206, [ ], 3306, [ ], 3406

115, 315, [ ], [ ], 915   5100, 6100, [ ], [ ], 9100

3809, 3810, 3811, [ ], [ ]   4020, 4130, 4240, [ ], [ ]

Time spent: _____ min _____ sec. Total: _____ out of 32

STEP 1 (1 min) Warm-up

Start the timer

Answer these.

$400 + 300 =$ ☐ $250 + 500 =$ ☐ $700 + 290 =$ ☐ $180 + 320 =$ ☐

$680 + 240 =$ ☐ $270 + 440 =$ ☐ $234 + 400 =$ ☐ $500 + 179 =$ ☐

STEP 2 (2.5 min) Rapid calculation

Start the timer

Answer these.

$480 + 235 =$ ☐ $250 + 480 =$ ☐ $690 + 246 =$ ☐

$553 + 372 =$ ☐ $364 + 324 =$ ☐ $437 + 252 =$ ☐

$625 + 253 =$ ☐ $177 + 422 =$ ☐ $851 + 147 =$ ☐

$274 + 323 =$ ☐ $145 + 641 =$ ☐ $157 + 342 =$ ☐

$444 + 238 =$ ☐ $545 + 383 =$ ☐ $627 + 282 =$ ☐

$723 + 247 =$ ☐ $457 + 345 =$ ☐ $207 + 579 =$ ☐

$776 + 188 =$ ☐ $856 + 197 =$ ☐

STEP 3 (1.5 min) Challenge

Start the timer

Answer these.

$456 + 795 =$ ☐ $567 + 687 =$ ☐ $777 + 576 =$ ☐ $429 + 567 =$ ☐

$549 + 676 =$ ☐ $294 + 998 =$ ☐ $848 + 867 =$ ☐ $968 + 966 =$ ☐

5 Subtraction with three-digit numbers

Date: _____

Day of Week: _____

STEP 1 (1 min) **Warm-up**

Start the timer

Answer these.

63 – 28 = ☐   54 – 28 = ☐   72 – 27 = ☐   65 – 29 = ☐

77 – 58 = ☐   65 – 26 = ☐   82 – 53 = ☐   78 – 39 = ☐

91 – 46 = ☐   86 – 67 = ☐

STEP 2 (2.5 min) **Rapid calculation**

Start the timer

Answer these.

455 – 233 = ☐   566 – 444 = ☐   435 – 122 = ☐   656 – 413 = ☐

545 – 114 = ☐   734 – 231 = ☐   748 – 326 = ☐   976 – 754 = ☐

576 – 238 = ☐   546 – 227 = ☐   876 – 649 = ☐   663 – 248 = ☐

746 – 288 = ☐   567 – 439 = ☐   677 – 398 = ☐   954 – 375 = ☐

323 – 182 = ☐   711 – 526 = ☐

STEP 3 (1.5 min) **Challenge**

Start the timer

Fill in the boxes with >, < or =.

565 – 386 ☐ 180   725 – 347 ☐ 377   676 – 188 ☐ 488

456 – 287 ☐ 170   911 – 465 ☐ 445   823 – 546 ☐ 277

936 – 778 ☐ 157   655 – 266 ☐ 390   564 – 289 ☐ 274

Time spent: _____ min _____ sec. Total: _____ out of 37

STEP 1 ⏱ 1 min **Warm-up**

Start the timer

Answer these.

410 − 200 = ☐   220 + 100 = ☐   650 + 300 = ☐   520 + 400 = ☐

290 + 500 = ☐   780 + 200 = ☐   390 − 100 = ☐   420 − 300 = ☐

640 − 400 = ☐   980 − 500 = ☐

STEP 2 ⏱ 2.5 min **Rapid calculation**

Start the timer

Answer these.

417 − 300 = ☐   559 + 300 = ☐   620 + 209 = ☐   360 − 150 = ☐

865 − 260 = ☐   785 − 280 = ☐   145 + 150 = ☐   450 + 160 = ☐

125 + 270 = ☐   470 − 330 = ☐   572 − 170 = ☐   656 + 240 = ☐

712 + 180 = ☐   220 + 452 = ☐   1000 − 500 = ☐   624 − 448 = ☐

STEP 3 ⏱ 1.5 min **Challenge**

Start the timer

Fill in the missing numbers.

280 + ☐ = 490   220 + ☐ = 795   345 + ☐ = 565

306 − ☐ = 140   ☐ − 270 = 400   540 + ☐ = 880

895 − ☐ = 350   ☐ − 320 = 550   ☐ + 120 = 695

7 Estimating additions and subtractions with four-digit numbers

Date: _____

Day of Week: _____

STEP 1 (1 min) Warm-up

Start the timer

Write these numbers to the nearest hundreds.

4582 [ ]   3367 [ ]   5876 [ ]   9050 [ ]   6911 [ ]

8436 [ ]   5555 [ ]   1735 [ ]   4382 [ ]   6450 [ ]

STEP 2 (2.5 min) Rapid calculation

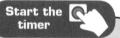
Start the timer

Estimate and then calculate.

Question	Estimate	Calculated answer
4374 + 2318 =		
1458 + 7136 =		
5436 + 2357 =		
4835 + 3197 =		

Question	Estimate	Calculated answer
5469 – 2587 =		
7437 – 3368 =		
8435 – 2157 =		
4354 – 1177 =		

STEP 3 (1.5 min) Challenge

Start the timer

Estimate and then calculate.

Question	Estimate	Calculated answer
7353 – 2578 =		
5478 + 3784 =		
9467 – 6688 =		

Question	Estimate	Calculated answer
3285 + 4986 =		
8844 – 3967 =		
4378 + 4884 =		

Time spent: _____ min _____ sec. Total: _____ out of 24

Date: _____

Day of Week: _____

STEP 1 (1 min) Warm-up

Start the timer

Answer these.

300 + 200 = ☐   500 + 4000 = ☐   2000 + 4000 = ☐

800 + 500 = ☐   3000 + 900 = ☐   4000 + 5000 = ☐

3000 + 6000 = ☐   1000 + 7000 = ☐   2000 + 7000 = ☐

STEP 2 (2.5 min) Rapid calculation

Start the timer

Answer these.

3200 + 1700 = ☐   2800 + 1100 = ☐   1300 + 7100 = ☐

6200 + 2700 = ☐   5500 + 4200 = ☐   4300 + 2500 = ☐

5300 + 2400 = ☐   4500 + 5400 = ☐   3100 + 2800 = ☐

3700 + 2100 = ☐   1200 + 2400 = ☐   6500 + 2300 = ☐

3200 + 3600 = ☐   7200 + 2200 = ☐   6300 + 2400 = ☐

STEP 3 (1.5 min) Challenge

Start the timer

Answer these.

1203 + 2203 = ☐   2304 + 3440 = ☐   4102 + 2475 = ☐

3102 + 2457 = ☐   3540 + 2321 = ☐   4362 + 1537 = ☐

5903 + 3024 = ☐   1004 + 3024 = ☐

Date: _____

Day of Week: _____

STEP 1 (1 min) Warm-up

Start the timer

Answer these.

400 – 200 = [ ]   900 – 50 = [ ]   800 – 600 = [ ]   600 – 300 = [ ]

700 – 200 = [ ]   1100 – 500 = [ ]   500 – 400 = [ ]   700 – 400 = [ ]

400 – 100 = [ ]   900 – 300 = [ ]

STEP 2 (2.5 min) Rapid calculation

Start the timer

Answer these.

4200 – 2100 = [ ]   8700 – 5200 = [ ]   7400 – 3300 = [ ]

7600 – 4300 = [ ]   6700 – 5200 = [ ]   7300 – 4300 = [ ]

9500 – 4100 = [ ]   5800 – 2600 = [ ]   8400 – 2400 = [ ]

4800 – 2400 = [ ]   5800 – 3200 = [ ]   6900 – 3600 = [ ]

6600 – 4400 = [ ]   7900 – 3500 = [ ]   5900 – 4100 = [ ]

STEP 3 (1.5 min) Challenge

Start the timer

Answer these.

5302 – 2201 = [ ]   7450 – 2320 = [ ]   5479 – 2361 = [ ]

4090 – 2030 = [ ]   7489 – 3265 = [ ]   6590 – 4560 = [ ]

4807 – 2405 = [ ]   5784 – 2561 = [ ]

Time spent: _____ min _____ sec. Total: _____ out of 33

Date: _____

Day of Week: _____

STEP 1 (1 min) Warm-up

Answer these.

$28 + 52 = $ ☐

$80 - 52 = $ ☐

$80 - 28 = $ ☐

$91 - 37 = $ ☐

$54 + 37 = $ ☐

$91 - 54 = $ ☐

$260 + 530 = $ ☐

$790 - 260 = $ ☐

$790 - 530 = $ ☐

$1200 - 400 = $ ☐

STEP 2 (2.5 min) Rapid calculation

 TIP *Remember the inverse relationship between addition and subtraction.*

Fill in the missing numbers.

☐ $+ 12 = 60$

$72 + $ ☐ $= 100$

☐ $+ 45 = 155$

$347 - $ ☐ $= 227$

$1000 - $ ☐ $= 230$

$125 + $ ☐ $= 200$

☐ $- 39 = 46$

☐ $- 81 = 81$

$93 + $ ☐ $= 153$

☐ $- 430 = 250$

☐ $- 310 = 210$

$112 + $ ☐ $= 160$

☐ $+ 70 = 320$

☐ $- 350 = 650$

$729 - $ ☐ $= 420$

STEP 3 (1.5 min) Challenge

Fill in the missing numbers.

☐ $- 64 = 36$

$372 - $ ☐ $= 122$

☐ $+ 180 = 340$

☐ $- 50 = 5 \times 24$

$402 + $ ☐ $- 224 = 200$

☐ $+ 75 = 5 \times 48$

☐ $- 650 = 200 + 150$

$81 - $ ☐ $= 119 - 59$

©HarperCollins*Publishers* 2019

Time spent: _____ min _____ sec. Total: _____ out of 33

13

Date: _____

Day of Week: _____

STEP 1 (1 min) **Warm-up**

Start the timer

Write the value of each Roman numeral.

I = [ ]   V = [ ]   X = [ ]   L = [ ]

C = [ ]   D = [ ]   M = [ ]

STEP 2 (2.5 min) **Rapid calculation**

Start the timer

Write each Roman numeral as an Arabic numeral.

VIII = [ ]   IV = [ ]   XII = [ ]

XV = [ ]   XLIII = [ ]   LXVI = [ ]

IX = [ ]   LXXXII = [ ]   XXXVII = [ ]

STEP 3 (1.5 min) **Challenge**

Start the timer

Fill in the boxes using >, < or =.

VII [ ] 6   XIX [ ] 19   XXVIII [ ] 29

XXXIII [ ] 32   LXXIV [ ] 75   XCVI [ ] 96

Time spent: _____ min _____ sec. Total: _____ out of 22

Date: _____

Day of Week: _____

STEP 1 (1 min) Warm-up

Start the timer

Answer these.

$0 \div 5 =$ ☐ $\qquad$ $6 \times 0 =$ ☐ $\qquad$ $64 \div 8 =$ ☐ $\qquad$ $55 + 24 =$ ☐

$77 \div 7 =$ ☐ $\qquad$ $36 - 26 =$ ☐ $\qquad$ $18 \div 9 =$ ☐ $\qquad$ $1 \times 6 =$ ☐

$36 \div 3 =$ ☐ $\qquad$ $12 \times 5 =$ ☐ $\qquad$ $29 + 34 =$ ☐ $\qquad$ $6 \times 8 =$ ☐

STEP 2 (2.5 min) Rapid calculation

Start the timer

Answer these.

$2 \times 6 =$ ☐ $\qquad$ $8 \times 6 =$ ☐ $\qquad$ $6 \times 3 =$ ☐ $\qquad$ $11 \times 6 =$ ☐

$6 \times 5 =$ ☐ $\qquad$ $6 \times 12 =$ ☐ $\qquad$ $6 \times 7 =$ ☐ $\qquad$ $9 \times 6 =$ ☐

$24 \div 6 =$ ☐ $\qquad$ $30 \div 6 =$ ☐ $\qquad$ $18 \div 6 =$ ☐ $\qquad$ $54 \div 6 =$ ☐

$48 \div 6 =$ ☐ $\qquad$ $10 \times 6 =$ ☐ $\qquad$ $6 \div 6 =$ ☐ $\qquad$ $6 \times 9 =$ ☐

$66 \div 6 =$ ☐ $\qquad$ $6 \div 1 =$ ☐ $\qquad$ $6 \times 6 =$ ☐ $\qquad$ $60 \div 6 =$ ☐

STEP 3 (1.5 min) Challenge

Start the timer

Answer these.

$6 \times 7 + 3 =$ ☐ $\qquad$ $9 \times 6 + 6 =$ ☐ $\qquad$ $6 \times 10 + 14 =$ ☐

$8 \times 6 + 11 =$ ☐ $\qquad$ $12 \times 6 + 18 =$ ☐ $\qquad$ $36 \div 6 + 5 =$ ☐

$30 \div 10 + 16 =$ ☐ $\qquad$ $18 \div 6 + 72 =$ ☐ $\qquad$ $72 \div 6 + 81 =$ ☐

Date: _____

Day of Week: _____

STEP 1 (1 min) Warm-up

Start the timer

Answer these.

4 × 8 = ☐   26 + 18 = ☐   72 ÷ 9 = ☐   56 − 28 = ☐

45 ÷ 5 = ☐   72 − 28 = ☐   5 × 6 = ☐   34 + 29 = ☐

100 − 46 = ☐   4 × 12 = ☐   62 + 9 = ☐   12 × 2 = ☐

STEP 2 (2.5 min) Rapid calculation

Start the timer

Answer these.

7 × 2 = ☐   4 × 7 = ☐   8 × 7 = ☐   5 × 7 = ☐

7 × 6 = ☐   2 × 7 = ☐   7 × 3 = ☐   7 × 4 = ☐

7 × 5 = ☐   7 × 12 = ☐   7 × 8 = ☐   7 × 9 = ☐

1 × 7 = ☐   10 × 7 = ☐   3 × 7 = ☐   0 × 7 = ☐

7 × 11 = ☐   6 × 7 = ☐   12 × 7 = ☐   7 × 7 = ☐

STEP 3 (1.5 min) Challenge

Start the timer

Answer these.

21 ÷ 7 = ☐   35 ÷ 7 = ☐   49 ÷ 7 = ☐

14 ÷ 7 = ☐   56 ÷ 7 = ☐   77 ÷ 7 = ☐

84 ÷ 7 = ☐   42 ÷ 7 = ☐   63 ÷ 7 = ☐

Time spent: _____ min _____ sec. Total: _____ out of 41

Date: _____

Day of Week: _____

STEP 1 (1 min) Warm-up

Start the timer

Answer these.

3 × 11 = ☐   63 – 45 = ☐   10 × 0 = ☐   39 + 42 = ☐

7 × 12 = ☐   12 × 6 = ☐   77 ÷ 7 = ☐   2 × 11 = ☐

90 – 69 = ☐   0 ÷ 6 = ☐   42 – 14 = ☐   36 ÷ 3 = ☐

STEP 2 (2.5 min) Rapid calculation

Start the timer

Answer these.

2 × 9 = ☐   5 × 9 = ☐   9 × 9 = ☐   9 × 10 = ☐

9 × 6 = ☐   11 × 9 = ☐   7 × 9 = ☐   9 × 8 = ☐

4 × 9 = ☐   9 × 12 = ☐   0 × 9 = ☐   54 ÷ 9 = ☐

18 ÷ 9 = ☐   108 ÷ 9 = ☐   81 ÷ 9 = ☐   99 ÷ 9 = ☐

63 ÷ 9 = ☐   27 ÷ 9 = ☐   45 ÷ 9 = ☐   90 ÷ 9 = ☐

STEP 3 (1.5 min) Challenge

Start the timer

Fill in the missing numbers.

☐ ÷ 9 = 3   36 ÷ ☐ = 4   ☐ × 9 = 45

☐ ÷ 9 = 7   ☐ × 9 = 81   ☐ ÷ 9 = 10

9 × ☐ = 6 × 6   81 ÷ ☐ = 27 ÷ 3   ☐ × 9 = 108 ÷ 2

5 threes plus 3 threes equals 8 threes

Date: _____

Day of Week: _____

STEP 1 (1 min) Warm-up

Start the timer

Answer these.

34 + 21 = ☐   27 − 13 = ☐   41 + 28 = ☐   38 − 18 = ☐

16 + 34 = ☐   30 + 58 = ☐   56 − 39 = ☐   75 − 38 = ☐

56 + 44 = ☐   88 − 59 = ☐   63 − 36 = ☐   37 + 47 = ☐

STEP 2 (2.5 min) Rapid calculation

Start the timer

1. Fill in the missing numbers.

$5 \times 2 + 3 \times 2 =$ ☐ $\times 2 =$ ☐ $3 \times 3 + 6 \times 3 =$ ☐ $\times 3 =$ ☐

$6 \times 4 + 4 \times 4 =$ ☐ $\times 4 =$ ☐ $4 \times 5 + 2 \times 5 =$ ☐ $\times 5 =$ ☐

$6 \times 6 + 3 \times 6 =$ ☐ $\times 6 =$ ☐ $4 \times 7 + 2 \times 7 =$ ☐ $\times 7 =$ ☐

2. Answer these.

$3 \times 8 + 4 \times 8 =$ ☐ $6 \times 9 + 2 \times 9 =$ ☐

$5 \times 7 + 2 \times 7 =$ ☐ $5 \times 2 + 6 \times 2 =$ ☐

$7 \times 5 + 5 \times 5 =$ ☐ $4 \times 4 + 4 \times 8 =$ ☐

STEP 3 (1.5 min) Challenge

Start the timer

Fill in the missing numbers.

$19 \times 4 =$ ☐ $\times$ ☐ $+$ ☐ $\times$ ☐ $=$ ☐ $+$ ☐ $=$ ☐

$12 \times 8 =$ ☐ $\times$ ☐ $+$ ☐ $\times$ ☐ $=$ ☐ $+$ ☐ $=$ ☐

$13 \times 4 =$ ☐ $\times$ ☐ $+$ ☐ $\times$ ☐ $=$ ☐ $+$ ☐ $=$ ☐

$15 \times 6 =$ ☐ $\times$ ☐ $+$ ☐ $\times$ ☐ $=$ ☐ $+$ ☐ $=$ ☐

Time spent: _____ min _____ sec. Total: _____ out of 28

Date: _____

Day of Week: _____

STEP 1 (1 min) Warm-up

Start the timer

Answer these.

172 – 72 = ☐   2 × 85 = ☐   270 + 55 = ☐   387 ÷ 3 = ☐

275 × 5 = ☐   510 ÷ 3 = ☐   4 × 29 = ☐   123 × 3 = ☐

844 ÷ 4 = ☐   130 × 4 = ☐

STEP 2 (2.5 min) Rapid calculation

Start the timer

Answer these.

45 × 5 = ☐   223 × 3 = ☐   60 × 13 = ☐

41 × 8 = ☐   37 × 8 = ☐   74 × 3 = ☐

340 × 4 = ☐   39 × 4 = ☐   54 × 12 = ☐

46 × 7 = ☐   62 × 9 = ☐   78 × 3 = ☐

28 × 5 = ☐   62 × 8 = ☐   83 × 8 = ☐

STEP 3 (1.5 min) Challenge

Start the timer

Answer these.

420 × 5 = ☐   940 ÷ 4 = ☐

728 ÷ 8 = ☐   162 × 5 = ☐

13 × 17 = ☐   840 ÷ 6 = ☐

235 ÷ 5 = ☐   56 × 6 = ☐

Time spent: _____ min _____ sec. Total: _____ out of 33

Date: _____

Day of Week: _____

STEP 1 (1 min) Warm-up

Start the timer

Answer these.

$624 \div 6 =$ ☐

$618 \div 6 =$ ☐

$936 \div 3 =$ ☐

$256 \div 8 =$ ☐

$270 \div 6 =$ ☐

$345 \div 5 =$ ☐

$161 \div 7 =$ ☐

$768 \div 4 =$ ☐

STEP 2 (2.5 min) Rapid calculation

Start the timer

Answer these.

$369 \div 3 =$ ☐

$864 \div 6 =$ ☐

$405 \div 5 =$ ☐

$700 \div 5 =$ ☐

$356 \div 4 =$ ☐

$468 \div 2 =$ ☐

$645 \div 5 =$ ☐

$133 \div 7 =$ ☐

$330 \div 6 =$ ☐

$207 \div 3 =$ ☐

$752 \div 8 =$ ☐

$954 \div 6 =$ ☐

$963 \div 9 =$ ☐

$378 \div 7 =$ ☐

$504 \div 9 =$ ☐

STEP 3 (1.5 min) Challenge

Start the timer

Answer these.

$104 \div 4 =$ ☐

$272 \div 8 =$ ☐

$756 \div 9 =$ ☐

$324 \div 3 =$ ☐

$666 \div 6 =$ ☐

$567 \div 7 =$ ☐

$150 \div 6 =$ ☐

$465 \div 5 =$ ☐

$900 \div 9 =$ ☐

Time spent: _____ min _____ sec. Total: _____ out of 32

Date: _____

Day of Week: _____

STEP 1 (1 min) **Warm-up**

Start the timer

Answer these.

30 × 30 = [　　]   70 × 60 = [　　]   50 × 40 = [　　]   20 × 30 = [　　]

60 × 80 = [　　]   60 × 90 = [　　]   90 × 30 = [　　]   20 × 70 = [　　]

40 × 80 = [　　]   50 × 50 = [　　]

STEP 2 (2.5 min) **Rapid calculation**

Start the timer

Answer these.

16 × 30 = [　　]   47 × 20 = [　　]   12 × 40 = [　　]

19 × 30 = [　　]   25 × 40 = [　　]   13 × 20 = [　　]

24 × 40 = [　　]   22 × 50 = [　　]   80 × 12 = [　　]

70 × 13 = [　　]   30 × 25 = [　　]   90 × 11 = [　　]

24 × 30 = [　　]   30 × 17 = [　　]   20 × 48 = [　　]

STEP 3 (1.5 min) **Challenge**

Start the timer

Answer these.

20 × 40 = [　　]   25 × 20 = [　　]   15 × 30 = [　　]

13 × 40 = [　　]   16 × 90 = [　　]   16 × 60 = [　　]

17 × 40 = [　　]   16 × 80 = [　　]   15 × 70 = [　　]

---

19 Multiplying a two-digit number by a two-digit number

Date: _____

Day of Week: _____

STEP 1 (1 min) Warm-up

Start the timer

Answer these.

1. $12 \times 2 =$ ☐

   $12 \times 20 =$ ☐

2. $24 \times 3 =$ ☐

   $24 \times 30 =$ ☐

3. $13 \times 2 =$ ☐

   $13 \times 20 =$ ☐

4. $32 \times 5 =$ ☐

   $32 \times 50 =$ ☐

5. $29 \times 7 =$ ☐

   $29 \times 70 =$ ☐

STEP 2 (2.5 min) Rapid calculation

Start the timer

Answer these. The first one has been done for you.

1. 25×12

   $= 25 \times \boxed{10} + 25 \times \boxed{2}$

   $= \boxed{250} + \boxed{50}$

   $= \boxed{300}$

2. 49×23

   $= 49 \times \boxed{} + 49 \times \boxed{}$

   $= \boxed{} + \boxed{}$

   $= \boxed{}$

3. 35×46

   $= \boxed{} \times \boxed{} + \boxed{} \times \boxed{}$

   $= \boxed{} + \boxed{}$

   $= \boxed{}$

4. 21×18

   $= \boxed{} \times \boxed{} + \boxed{} \times \boxed{}$

   $= \boxed{} + \boxed{}$

   $= \boxed{}$

5. 29×52

   $= \boxed{} \times \boxed{} + \boxed{} \times \boxed{}$

   $= \boxed{} + \boxed{}$

   $= \boxed{}$

6. 71×34

   $= \boxed{} \times \boxed{} + \boxed{} \times \boxed{}$

   $= \boxed{} + \boxed{}$

   $= \boxed{}$

STEP 3 (1.5 min) Challenge

Start the timer

Answer these.

$15 \times 11 =$ ☐ $25 \times 11 =$ ☐ $11 \times 38 =$ ☐ $11 \times 81 =$ ☐

$11 \times 34 =$ ☐ $14 \times 11 =$ ☐ $11 \times 63 =$ ☐ $11 \times 92 =$ ☐

Time spent: _____ min _____ sec. Total: _____ out of 23

Date: _____

Day of Week: _____

STEP 1 (1 min) Warm-up

Start the timer

Answer these.

1. $15 \times 3 =$ ☐

$150 \times 3 =$ ☐

2. $35 \times 4 =$ ☐

$350 \times 4 =$ ☐

3. $16 \times 8 =$ ☐

$160 \times 8 =$ ☐

4. $36 \times 6 =$ ☐

$360 \times 6 =$ ☐

5. $27 \times 4 =$ ☐

$270 \times 4 =$ ☐

STEP 2 (2.5 min) Rapid calculation

Start the timer

Answer these.

$320 \times 3 =$ ☐

$430 \times 3 =$ ☐

$150 \times 8 =$ ☐

$640 \times 5 =$ ☐

$140 \times 8 =$ ☐

$280 \times 4 =$ ☐

$540 \times 4 =$ ☐

$440 \times 5 =$ ☐

$270 \times 7 =$ ☐

$210 \times 9 =$ ☐

$190 \times 6 =$ ☐

$630 \times 5 =$ ☐

$370 \times 3 =$ ☐

$260 \times 6 =$ ☐

$510 \times 8 =$ ☐

STEP 3 (1.5 min) Challenge

Start the timer

Answer these.

$330 \times 7 =$ ☐

$290 \times 3 =$ ☐

$670 \times 5 =$ ☐

$860 \times 8 =$ ☐

$410 \times 5 =$ ☐

$560 \times 4 =$ ☐

$230 \times 9 =$ ☐

$480 \times 8 =$ ☐

©HarperCollins*Publishers* 2019

Time spent: _____ min _____ sec. Total: _____ out of 33

23

Date: _____

Day of Week: _____

STEP 1 (1 min) **Warm-up**

Start the timer

Answer these.

400 ÷ 2 = ☐   200 ÷ 5 = ☐   600 ÷ 5 = ☐   300 ÷ 6 = ☐

810 ÷ 9 = ☐   420 ÷ 7 = ☐   320 ÷ 4 = ☐   640 ÷ 8 = ☐

540 ÷ 9 = ☐   120 ÷ 4 = ☐

STEP 2 (2.5 min) **Rapid calculation**

Start the timer

Answer these.

400 ÷ 20 = ☐   300 ÷ 60 = ☐   500 ÷ 50 = ☐

400 ÷ 80 = ☐   600 ÷ 10 = ☐   120 ÷ 20 = ☐

240 ÷ 30 = ☐   480 ÷ 60 = ☐   560 ÷ 80 = ☐

350 ÷ 70 = ☐   810 ÷ 90 = ☐   360 ÷ 60 = ☐

540 ÷ 60 = ☐   390 ÷ 30 = ☐   280 ÷ 70 = ☐

STEP 3 (1.5 min) **Challenge**

Start the timer

Answer these.

720 ÷ 80 = ☐   180 ÷ 20 = ☐

420 ÷ 60 = ☐   280 ÷ 20 = ☐

640 ÷ 80 = ☐   490 ÷ 70 = ☐

300 ÷ 50 = ☐   840 ÷ 20 = ☐

Time spent: _____ min _____ sec. Total: _____ out of 33

STEP 1 (1 min) Warm-up

Start the timer

Answer these.

4 × 53 = [ ]   37 × 6 = [ ]   48 × 4 = [ ]   8 × 22 = [ ]

61 × 7 = [ ]   5 × 76 = [ ]   39 × 6 = [ ]   7 × 64 = [ ]

STEP 2 (2.5 min) Rapid calculation

Start the timer

Estimate first, then use the expanded written method to work out the answers. Remember to use your estimate to check your answer. The first one has been done for you.

1. 76 × 3 = **228**

Estimate: **240**

```
      7   6
  ×       3
 ─────────────
      1   8
  2   1   0
 ─────────────
  2   2   8
```

2. 87 × 6 =

Estimate:

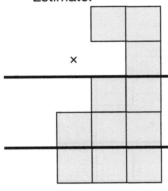

3. 94 × 8 =

Estimate:

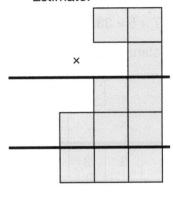

4. 317 × 4 =

Estimate:

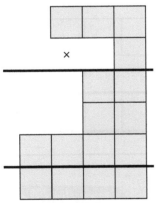

5. 456 × 3 =

Estimate:

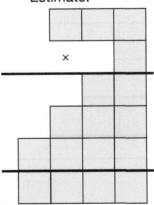

6. 748 × 5 =

Estimate:

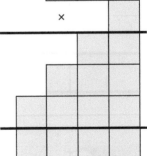

STEP 3 (1.5 min) Challenge

Start the timer

Use the expanded written method to answer these. Use a separate sheet of paper for your working.

28 × 6 = [ ]   7 × 37 = [ ]   53 × 8 = [ ]

4 × 426 = [ ]   578 × 6 = [ ]   8 × 664 = [ ]

Date: _____

Day of Week: _____

STEP 1 (1 min) **Warm-up**

Start the timer

Answer these.

$3 \times 78 =$ ☐ $84 \times 5 =$ ☐ $66 \times 7 =$ ☐ $6 \times 92 =$ ☐

$87 \times 9 =$ ☐ $8 \times 68 =$ ☐ $93 \times 7 =$ ☐ $6 \times 83 =$ ☐

STEP 2 (2.5 min) **Rapid calculation**

Start the timer

Estimate first, then use short multiplication to work out the answers. Remember to use your estimate to check your answer. The first one has been done for you.

1. $67 \times 5 =$ **335**

Estimate: **350**

	6	7
×		5
3	3	5
	3	

2. $56 \times 9 =$

Estimate:

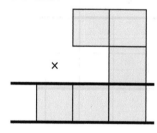

3. $93 \times 8 =$

Estimate:

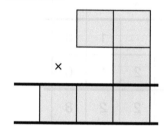

4. $247 \times 6 =$

Estimate:

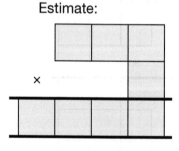

5. $486 \times 7 =$

Estimate:

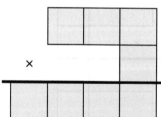

6. $728 \times 8 =$

Estimate:

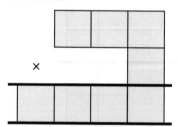

STEP 3 (1.5 min) **Challenge**

Start the timer

Use short multiplication to answer these. Use a separate sheet of paper for your working.

$59 \times 6 =$ ☐ $8 \times 73 =$ ☐ $88 \times 9 =$ ☐

$4 \times 358 =$ ☐ $678 \times 6 =$ ☐ $8 \times 894 =$ ☐

Time spent: _____ min _____ sec. Total: _____ out of 19

Date: _____

Day of Week: _____

STEP 1 (1 min) **Warm-up**

Start the timer

Answer these.

1. $60 \times 8 =$ ☐

$480 \div 60 =$ ☐

$480 \div 8 =$ ☐

2. $50 \times 6 =$ ☐

$300 \div 50 =$ ☐

$300 \div 6 =$ ☐

3. $14 \times 15 =$ ☐

$210 \div 14 =$ ☐

$210 \div 15 =$ ☐

4. $16 \times 60 =$ ☐

$960 \div 60 =$ ☐

$960 \div 16 =$ ☐

STEP 2 (2.5 min) **Rapid calculation**

Start the timer

 TIP *Remember the inverse relationship between multiplication and division.*

Fill in the missing numbers.

☐ $\div 12 = 8$

$100 \times$ ☐ $= 1000$

☐ $\div 86 = 30$

$165 \div$ ☐ $= 11$

$720 \div$ ☐ $= 30$

$95 \div$ ☐ $= 5$

☐ $\times 24 = 120$

$1000 \div$ ☐ $= 8$

☐ $\times 12 = 600$

$50 \times$ ☐ $= 900$

$360 \div$ ☐ $= 9$

☐ $\times 4 = 720$

☐ $\times 5 = 80$

☐ $\times 40 = 960$

$450 \div$ ☐ $= 30$

STEP 3 (1.5 min) **Challenge**

Start the timer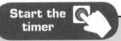

Fill in the missing numbers.

☐ $\times 12 = 200 + 40$

☐ $\times 8 = 360 \div 9$

☐ $\times 14 = 6 \times 7 \times 5$

$570 \div$ ☐ $= 5 \times 6$

☐ $\times 4 = 120 \times 2$

☐ $\div 25 = 5 \times 5$

$36 \times$ ☐ $= 72 \times 20$

☐ $\div 28 = 72 \div 9$

Time spent: _____ min _____ sec. Total: _____ out of 35

Date: _____

Day of Week: _____

STEP 1 (1 min) **Warm-up**

Start the timer

Write a fraction to describe the shaded part.

☐   ☐   ☐   ☐   ☐

STEP 2 (2.5 min) **Rapid calculation**

Start the timer

Fill in the missing numbers.

$\frac{1}{2}$ of 12 ☆s is ☐ ☆s. $\frac{1}{5}$ of 25 ☆s is ☐ ☆s. $\frac{1}{4}$ of 20 ◎s is ☐ ◎s.

$\frac{1}{2}$ of 10 ☐s is ☐ ☐s. $\frac{1}{7}$ of 14 △s is ☐ △s. $\frac{1}{4}$ of 16 ◇s is ☐ ◇s.

$\frac{1}{6}$ of 24 ◎s is ☐ ◎s. $\frac{1}{9}$ of 36 ■s is ☐ ■s. $\frac{1}{5}$ of 40 △s is ☐ △s.

STEP 3 (1.5 min) **Challenge**

Start the timer

Write **T** if the fraction matches the darker-shaded area of the shape or set, and **F** if it does not.

1.
$\frac{1}{4}$ ☐

2.
$\frac{1}{3}$ ☐

3.
$\frac{1}{2}$ ☐

4.
$\frac{1}{4}$ ☐

5.
5 m
5 m   5 m
$\frac{1}{2}$ ☐

6.
$\frac{1}{4}$ ☐

7.
$\frac{1}{3}$ ☐

8.
$\frac{1}{3}$ ☐

Time spent: _____ min _____ sec. Total: _____ out of 22

Date: _____

Day of Week: _____

STEP 1 (1 min) Warm-up

Fill in the missing numbers.

$\dfrac{3}{7} + \dfrac{2}{7} = \dfrac{\boxed{3} + \boxed{}}{\boxed{7}} = \dfrac{\boxed{}}{\boxed{7}}$

$\dfrac{1}{5} + \dfrac{3}{5} = \dfrac{\boxed{} + \boxed{}}{\boxed{}} = \dfrac{\boxed{}}{\boxed{}}$

$\dfrac{5}{10} + \dfrac{4}{10} = \dfrac{\boxed{} + \boxed{}}{\boxed{}} = \dfrac{\boxed{}}{\boxed{}}$

$\dfrac{13}{25} + \dfrac{11}{25} = \dfrac{\boxed{}}{\boxed{}} = \dfrac{\boxed{}}{\boxed{}}$

$\dfrac{2}{8} + \dfrac{5}{8} = \dfrac{\boxed{}}{\boxed{}} = \dfrac{\boxed{}}{\boxed{}}$

$\dfrac{28}{123} + \dfrac{49}{123} = \dfrac{\boxed{}}{\boxed{}} = \dfrac{\boxed{}}{\boxed{}}$

STEP 2 (2.5 min) Rapid calculation

 TIP *When adding or subtracting fractions with the same denominator, simply add or subtract the numerators and leave the denominator unchanged.*

Answer these.

$\dfrac{5}{9} + \dfrac{2}{9} = \boxed{}$ $\dfrac{5}{58} + \dfrac{34}{58} = \boxed{}$ $\dfrac{11}{20} + \dfrac{4}{20} = \boxed{}$ $\dfrac{3}{11} + \dfrac{5}{11} = \boxed{}$

$\dfrac{5}{12} + \dfrac{4}{12} = \boxed{}$ $\dfrac{14}{19} + \dfrac{2}{19} = \boxed{}$ $\dfrac{4}{15} + \dfrac{6}{15} = \boxed{}$ $\dfrac{8}{17} + \dfrac{7}{17} = \boxed{}$

$\dfrac{13}{47} + \dfrac{24}{47} = \boxed{}$ $\dfrac{3}{8} + \dfrac{4}{8} = \boxed{}$ $\dfrac{15}{39} + \dfrac{22}{39} = \boxed{}$ $\dfrac{11}{20} + \dfrac{2}{20} = \boxed{}$

$\dfrac{53}{108} + \dfrac{24}{108} = \boxed{}$ $\dfrac{28}{209} + \dfrac{12}{209} = \boxed{}$ $\dfrac{28}{107} + \dfrac{57}{107} = \boxed{}$

STEP 3 (1.5 min) Challenge

Fill in the missing fractions.

$\dfrac{2}{10} + \dfrac{3}{10} + \dfrac{4}{10} = \boxed{}$

$\dfrac{5}{17} + \dfrac{7}{17} + \dfrac{3}{17} = \boxed{}$

$\dfrac{25}{137} + \dfrac{36}{137} + \dfrac{40}{137} = \boxed{}$

$\dfrac{12}{29} + \dfrac{3}{29} + \dfrac{4}{29} = \boxed{}$

$\dfrac{5}{18} + \boxed{} = \dfrac{1}{2}$

$\dfrac{5}{24} + \boxed{} = \dfrac{1}{4}$

$\boxed{} + \dfrac{17}{225} = \dfrac{1}{5}$

$\boxed{} + \dfrac{7}{108} = \dfrac{1}{6}$

Time spent: _____ min _____ sec. Total: _____ out of 29

Date: _____

Day of Week: _____

STEP 1 (1 min) Warm-up

Start the timer

Answer these.

$\dfrac{2}{11} + \dfrac{7}{11} = \boxed{}$

$\dfrac{5}{13} + \dfrac{8}{13} = \boxed{}$

$\dfrac{3}{12} + \dfrac{7}{12} = \boxed{}$

$\dfrac{12}{19} + \dfrac{3}{19} = \boxed{}$

$\dfrac{12}{31} + \dfrac{17}{31} = \boxed{}$

$\dfrac{18}{20} - \dfrac{15}{20} = \boxed{}$

$1 - \dfrac{27}{41} = \boxed{}$

$\dfrac{25}{47} - \dfrac{25}{47} = \boxed{}$

$\dfrac{57}{64} - \dfrac{25}{64} = \boxed{}$

STEP 2 (2.5 min) Rapid calculation

Start the timer

Answer these.

$\dfrac{7}{16} + \dfrac{8}{16} = \boxed{}$

$1 - \dfrac{29}{40} = \boxed{}$

$\dfrac{16}{50} + \dfrac{34}{50} = \boxed{}$

$\dfrac{32}{65} + \dfrac{18}{65} = \boxed{}$

$\dfrac{17}{26} + \dfrac{8}{26} = \boxed{}$

$\dfrac{17}{29} + \dfrac{12}{29} = \boxed{}$

$\dfrac{32}{45} - \dfrac{15}{45} = \boxed{}$

$\dfrac{25}{34} + \dfrac{9}{34} = \boxed{}$

$\dfrac{29}{37} - \dfrac{14}{37} = \boxed{}$

$1 - \dfrac{29}{37} = \boxed{}$

$1 - \dfrac{17}{30} = \boxed{}$

$\dfrac{63}{74} - \dfrac{38}{74} = \boxed{}$

$\dfrac{27}{94} + \dfrac{35}{94} = \boxed{}$

$\dfrac{23}{29} - \dfrac{15}{29} = \boxed{}$

$\dfrac{32}{43} - \dfrac{17}{43} = \boxed{}$

$1 - \dfrac{11}{18} = \boxed{}$

STEP 3 (1.5 min) Challenge

Start the timer

Fill in the missing fractions.

$\left(1 - \dfrac{7}{16}\right) + \dfrac{8}{16} = \boxed{}$

$\dfrac{30}{74} + \dfrac{44}{74} - \dfrac{18}{91} = \boxed{}$

$\dfrac{32}{45} - \dfrac{11}{45} - \dfrac{2}{45} = \boxed{}$

$\dfrac{24}{93} + \dfrac{23}{93} + \dfrac{14}{93} = \boxed{}$

$\dfrac{32}{45} - \boxed{} = \dfrac{1}{5}$

$\dfrac{48}{96} - \boxed{} = \dfrac{1}{12}$

$\boxed{} - \dfrac{3}{84} = \dfrac{1}{21}$

$\boxed{} - \dfrac{12}{44} = \dfrac{1}{4}$

Time spent: _____ min _____ sec. Total: _____ out of 33

©HarperCollins*Publishers* 2019

STEP 1 (1 min) Warm-up

Circle the numbers which have decimal places.

0.15   7.23   64   100.04   102   0.3   117   20.1   38   0.04   30.2   33

STEP 2 (2.5 min) Rapid calculation

Complete these sentences. The first one has been done for you.

0.08 m means [8] centimetres, read as <u>zero point zero eight</u> metres.

1.50 m means [ ] metre [ ] centimetres, read as .... metres.

4.25 m means [ ] metres [ ] centimetres, read as .... metres.

1.20 m means [ ] metre [ ] centimetres, read as .... metres.

2.18 m means [ ] metres [ ] centimetres, read as .... metres.

49.90 m means [ ] metres [ ] centimetres, read as .... metres.

STEP 3 (1.5 min) Challenge

 TIP $1\,m^2 = 10\,000\,cm^2$

Fill in the missing numbers.

2.2 kilograms means [ ] grams.

7.6 kilometres means [ ] metres.

£4.58 means [ ] pounds [ ] pence.

0.85 metres means [ ] centimetres.

1.6 tonnes means [ ] kilograms.

0.09 square metres means [ ] square centimetres.

Time spent: _____ min _____ sec. Total: _____ out of 18

Date: _____

Day of Week: _____

STEP 1 Warm-up

Start the timer

Represent the shaded parts as a fraction and a decimal.

1.

2.

3.

4.

□ □   □ □   □ □   □ □

STEP 2 Rapid calculation

Start the timer

Fill in the missing fractions and decimals.

1.
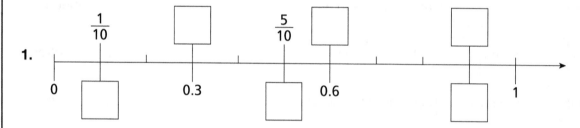

$\frac{1}{10}$ □ $\frac{5}{10}$ □ □

0 □ 0.3 □ 0.6 □ 1

2.
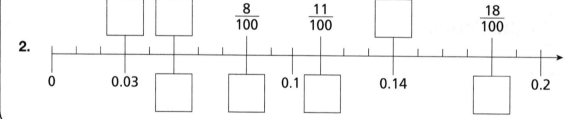

□ □ $\frac{8}{100}$ $\frac{11}{100}$ □ $\frac{18}{100}$

0 0.03 □ □ 0.1 □ 0.14 □ 0.2

STEP 3  Challenge

Start the timer

Answer these. Give your answers as decimals.

$\frac{64}{100} - \frac{35}{100} =$ □ $\frac{9}{10} - \frac{3}{10} =$ □

$\frac{220}{1000} + \frac{498}{1000} =$ □ $\frac{49}{200} + \frac{112}{200} =$ □

$\frac{85}{100} - \frac{75}{100} =$ □ $\frac{4}{10} + \frac{5}{10} =$ □

$\frac{745}{1000} - \frac{602}{1000} =$ □ $\frac{7}{100} + \frac{5}{100} =$ □

Time spent: _____ min _____ sec. Total: _____ out of 29

Date: _____

Day of Week: _____

STEP 1 (1 min) **Warm-up**

Start the timer

TIP *To compare two decimals, compare the whole parts first. The number with a greater whole part is greater. If the whole parts are the same, compare the tenths place, then the hundredths … until you find a greater digit with the same place value.*

Fill in the boxes with >, < or =.

0.87 ☐ 0.78   12.89 ☐ 12.98   5.67 ☐ 5.63   2.08 ☐ 4.23

7.08 ☐ 8.04   11.86 ☐ 10.86   4.2 ☐ 4.13   52.04 ☐ 52.43

STEP 2 (2.5 min) **Rapid calculation**

Start the timer

Fill in the boxes with >, < or =.

9.18 ☐ 8.19   3.19 ☐ 3.12   10.01 ☐ 10.0

5.47 ☐ 5.46   8.34 ☐ 8.34   9.27 ☐ 8.01

11.91 ☐ 12.01   1.14 ☐ 1.11   6.00 ☐ 6

6.0 ☐ 6.01   5.10 ☐ 5.1   1.25 ☐ 1.52

7.86 ☐ 7.68   0.35 ☐ 0.04   58.2 ☐ 59.2

4.1 ☐ 4.01   9.99 ☐ 9.9   7.77 ☐ 7.77

STEP 3 (1.5 min) **Challenge**

Start the timer

Fill in the boxes with >, < or =.

1.7 ☐ 1.07   2.4 ☐ 2.41   0.03 ☐ 0.13   4.67 ☐ 4.86

2.03 ☐ 2.04   1.7 ☐ 1.70   6.04 ☐ 6.040   7.87 ☐ 8.78

0.5 ☐ 0.05   1.0 ☐ 1

Date: _____

Day of Week: _____

STEP 1 (1 min) **Warm-up**

Start the timer

Write these fractions as decimals.

$\frac{7}{10} =$ ☐ $\frac{3}{100} =$ ☐ $\frac{38}{100} =$ ☐ $3\frac{92}{100} =$ ☐

$\frac{9}{100} =$ ☐ $\frac{21}{100} =$ ☐ $\frac{70}{100} =$ ☐ $\frac{5}{100} =$ ☐

$\frac{52}{100} =$ ☐ $5\frac{42}{100} =$ ☐

STEP 2 (2.5 min) **Rapid calculation**

Start the timer

Fill in the boxes with >, < or =.

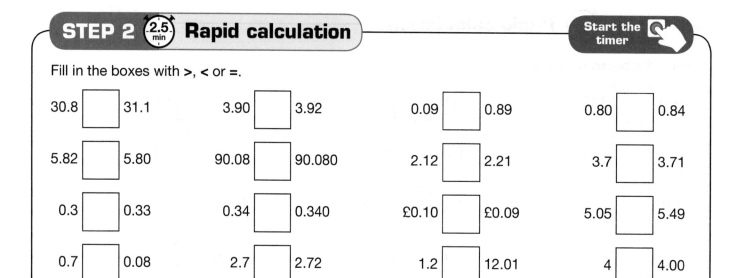

30.8 ☐ 31.1   3.90 ☐ 3.92   0.09 ☐ 0.89   0.80 ☐ 0.84

5.82 ☐ 5.80   90.08 ☐ 90.080   2.12 ☐ 2.21   3.7 ☐ 3.71

0.3 ☐ 0.33   0.34 ☐ 0.340   £0.10 ☐ £0.09   5.05 ☐ 5.49

0.7 ☐ 0.08   2.7 ☐ 2.72   1.2 ☐ 12.01   4 ☐ 4.00

STEP 3 (1.5 min) **Challenge**

Start the timer

Order these numbers from **least** to **greatest**.

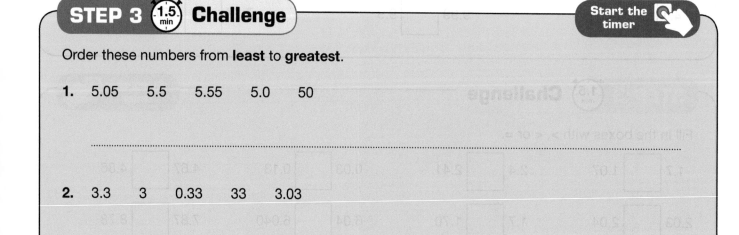

1. 5.05   5.5   5.55   5.0   50

....

2. 3.3   3   0.33   33   3.03

....

Time spent: _____ min _____ sec. Total: _____ out of 36

Date: _____

Day of Week: _____

STEP 1 (1 min) Warm-up

Start the timer

Write the number of each angle in the correct box.

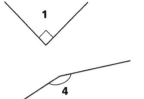

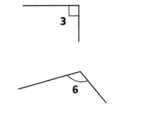
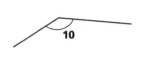

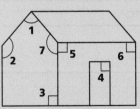

Right angles	Acute angles	Obtuse angles

STEP 2 (2.5 min) Rapid calculation

Start the timer

Which type of angle do the clock hands form at these times?

9:30   10:00   7:45   5:30   3:00   1:30   11:30   2:00   9:00

Write each time in the correct box.

Right angles	Acute angles	Obtuse angles

STEP 3 (1.5 min) Challenge

Start the timer

Which type of angles are shown in the diagram? The first one has been done for you.

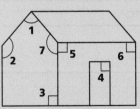

∠1: ......**acute**......angle   ∠2: .... angle   ∠3: .... angle

∠4: .... angle   ∠5: .... angle   ∠6: .... angle

∠7: .... angle

Time spent: _____ min _____ sec. Total: _____ out of 25

Date: _____

Day of Week: _____

STEP 1 (1 min) Warm-up

Start the timer

Complete these descriptions.

1. A shape enclosed by three .... is called a ....

2. A shape enclosed by .... line segments is called a quadrilateral.

3. A shape enclosed by five .... is called a ....

STEP 2 (2.5 min) Rapid calculation

Start the timer

Sort these shapes by writing their numbers in the correct boxes.

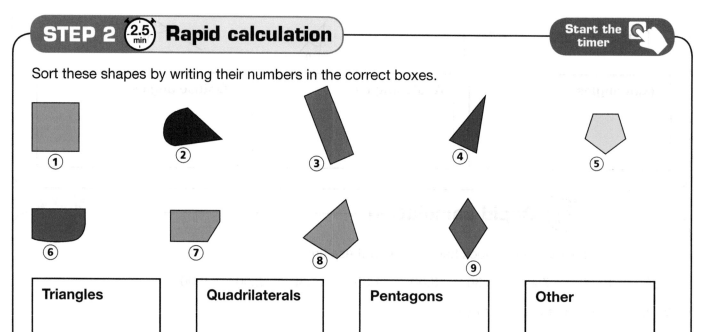

Triangles	Quadrilaterals	Pentagons	Other

STEP 3 (1.5 min) Challenge

Start the timer

Complete the table.

Shape	Number of sides	Number of angles	Name
△			
▱			
⏢			
⬠			

Time spent: _____ min _____ sec. Total: _____ out of 26

Date: _____

Day of Week: _____

STEP 1 (1 min) **Warm-up**
Start the timer

Complete these descriptions.

1. A triangle can be classified using its angles as a(n) .... triangle,

    a(n) .... triangle or a(n) .... triangle.

2. A triangle with an obtuse angle is called a(n) .... triangle.

3. A triangle with three acute angles is called a(n) .... triangle.

4. A triangle with a right angle is called a(n) .... triangle.

5. A triangle can be classified using its sides as a(n) .... triangle,

    a(n) .... triangle or a(n) .... triangle.

STEP 2 (2.5 min) **Rapid calculation**
Start the timer

Sort these triangles by writing their numbers in the correct box(es).

Acute	Obtuse	Right-angled	Isosceles	Equilateral

STEP 3 (1.5 min) **Challenge**
Start the timer

Complete these sentences.

1. A triangle has at least [ ] acute angle(s).

2. The side lengths of a triangle are 3 cm, 3 cm and 4 cm. According to its sides it is a(n)

    .... triangle.

3. A triangle with one axis of symmetry is called a(n) .... triangle.

    A triangle with three axes of symmetry is called a(n) .... triangle.

4. A triangle with three equal .... is called an equilateral triangle.

    It also has three equal ....

Time spent: _____ min _____ sec. Total: _____ out of 25

35 Line symmetry

Date: _____

Day of Week: _____

STEP 1 (1 min) Warm-up

Start the timer

Tick the boxes next to the symmetrical figures, then draw the lines of symmetry.

1. ☐

2. ☐

3. ☐

4. ☐

5. ☐

6. ☐

STEP 2 (2.5 min) Rapid calculation

Start the timer

Tick the boxes next to the figures with line symmetry, then draw the lines of symmetry.

1. ☐

2. ☐

3. ☐

4. ☐

5. 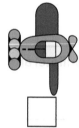 ☐

6. ☐

7. ☐

8. ☐

STEP 3 (1.5 min) Challenge

Start the timer

Complete these sentences.

1. A rectangle has ☐ lines of symmetry and a square has ☐ lines of symmetry.

2. An equilateral triangle has ☐ lines of symmetry.

3. An isosceles (but not equilateral) triangle has ☐ line(s) of symmetry.

4. A circle has .... lines of symmetry.

Time spent: _____ min _____ sec. Total: _____ out of 19   ©HarperCollinsPublishers 2019

Date: _____

Day of Week: _____

STEP 1 (1 min) Warm-up

Start the timer

Complete these.

1cm
1cm {

1. Area of a rectangle = ... × ....

   The area of the rectangle shown is [ ] cm².

   1cm
   } 1cm

2. Area of a square = ... × ....

   The area of the square shown is [ ] cm².

STEP 2 (2.5 min) Rapid calculation

Start the timer
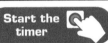

1. Work out the areas of these rectangles.

   length 7cm, width 5cm, area = [ ] cm²   length 20cm, width 15cm, area = [ ] cm²

   length 50cm, width 35cm, area = [ ] cm²   length 70m, width 42m, area = [ ] m²

   length 60m, width 45m, area = [ ] m²   length 40cm, width 20cm, area = [ ] cm²

2. Work out the areas of these squares.

   side length 15cm, area = [ ] cm²   side length 80cm, area = [ ] cm²

   side length 11m, area = [ ] m²   side length 20cm, area = [ ] cm²

STEP 3 (1.5 min) Challenge

Start the timer

1. Work out the areas of these rectangles.

   length 7m, width 50cm, area = [ ] cm☐   length 1m, width 32cm, area = [ ] cm☐

   length 2m, width 40cm, area = [ ] cm☐

2. Work out the areas of these squares.

   side length 5m, area = [ ] cm☐   side length 7m, area = [ ] cm☐

   side length 4m, area = [ ] cm☐

Time spent: _____ min _____ sec. Total: _____ out of 20

STEP 1 (1 min) Warm-up

Start the timer

Which is the most suitable metric unit? Fill in the abbreviations.

1. A baby giraffe is about 2 .... in height.

2. The distance from Mike's home to the library is 5 ....

3. For a running test in the PE class, every pupil has to run 100 ....

4. The full marathon distance is about 42 ....

5. A bicycle travels at around 300 .... a minute.

6. One lap of a running track is 400 m; an athlete runs five laps, which is 2 ....

7. A plane flies 900 .... an hour.

8. Mount Everest is about 8848 .... in height.

STEP 2 (2.5 min) Rapid calculation

Start the timer

Complete these conversions.

1 km = [____] m

4 km = [____] m

30 000 m = [____] km

9 km = [____] m

5000 m = [____] km

18 000 m = [____] km

7000 m = [____] km

80 km = [____] m

38 000 m = [____] km

19 km = [____] m

8 km 750 m = [____] m

7 km 7 m = [____] m

99 000 m = [____] km

16 km 40 m = [____] m

3 km 450 m = [____] m

STEP 3 (1.5 min) Challenge

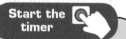

Start the timer

Fill in the missing numbers.

5 km 20 m = [____] m   9 km 67 m = [____] m   300 m + 4 km = [____] m

2 km + 5800 m = [____] m   8 km − 430 m = [____] m   8 km − 80 m = [____] m

30 km − [____] m = 600 m   [____] km − 1500 m = 500 m

Time spent: _____ min _____ sec. Total: _____ out of 31

Date: _____

Day of Week: _____

STEP 1 (1 min) Warm-up

Start the timer

1. The length of the outline of a plane figure is called the ....

2. Perimeter of a rectangle = ▢ × .... + ▢ × ....

3. Perimeter of a square = ▢ × ....

4. Work out the perimeter of each shape.

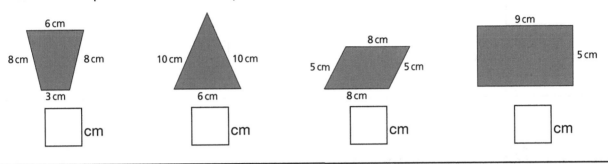

6 cm   8 cm   8 cm   3 cm   ▢ cm

10 cm   10 cm   6 cm   ▢ cm

8 cm   5 cm   5 cm   8 cm   ▢ cm

9 cm   5 cm   ▢ cm

STEP 2 (2.5 min) Rapid calculation

Start the timer

Complete the tables.

Rectangles	**Length**	18 cm	25 cm	40 cm	50 cm	100 cm
	Width	10 cm	20 cm	35 cm	42 cm	80 cm
	Perimeter					

Squares	**Side length**	34 cm	50 cm	72 cm	88 cm	43 cm
	Perimeter					

STEP 3 (1.5 min) Challenge

Start the timer

1. Find the missing side in each rectangle.

   perimeter 48 cm, length 8 cm, width ▢ cm

   perimeter 60 cm, width 4 cm, length ▢ cm

   perimeter 96 cm, length 36 cm, width ▢ cm

   perimeter 100 cm, width 18 cm, length ▢ cm

2. Find the missing side in each square.

   perimeter 76 cm, side length ▢ cm

   perimeter 92 cm, side length ▢ cm

   perimeter 64 cm, side length ▢ cm

   perimeter 48 cm, side length ▢ cm

Time spent: _____ min _____ sec. Total: _____ out of 25

Solving problems involving time and money

Date: _____

Day of Week: _____

STEP 1 (1 min) Warm-up

Start the timer

Fill in the missing numbers.

1 h = [ ] min   1 min = [ ] sec

1 day = [ ] h   1 week = [ ] days

1 year = [ ] months   1 h = [ ] sec

£1 = [ ] pence   2 min = [ ] sec

STEP 2 (2.5 min) Rapid calculation

Start the timer

Fill in the missing numbers.

4 h = [ ] min   3.5 min = [ ] sec   75 min = [ ] h

1.5 days = [ ] h $\frac{2}{5}$ h = [ ] min $\frac{2}{3}$ days = [ ] h

$\frac{1}{2}$ year = [ ] months $\frac{4}{5}$ of £1 = [ ] pence   £9.50 = [ ] pence

70 pence = £ [ ]   2 weeks = [ ] days   608 pence = £ [ ]

STEP 3 (1.5 min) Challenge

Start the timer

Fill in the missing numbers.

280 h = [ ] days [ ] h   590 pence = £ [ ] and [ ] pence

$\frac{3}{5}$ h = [ ] min = [ ] sec   [ ] h = 45 min = [ ] sec

4 weeks = [ ] days = [ ] h   2.5 years = [ ] months

Time spent: _____ min _____ sec. Total: _____ out of 29

Date: _____

Day of Week: _____

STEP 1 (1 min) Warm-up

Start the timer

Answer these.

600 ÷ 40 = ☐   180 + 52 = ☐   620 − 380 = ☐   800 ÷ 25 = ☐

99 × 7 = ☐   424 ÷ 4 = ☐   483 + 278 = ☐   37 × 40 = ☐

STEP 2 (2.5 min) Rapid calculation

Start the timer

Write a number sentence with brackets for each tree diagram. The first one has been done for you.

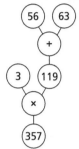 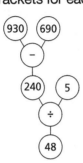 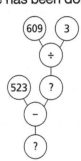

3 × (56 + 63) = 357

....

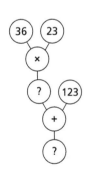 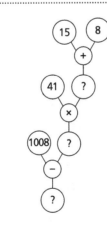 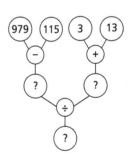

....

STEP 3 (1.5 min) Challenge

Start the timer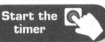

Answer these. Remember to complete the calculations in brackets first.

112 − (125 ÷ 5) = ☐   (228 + 129) ÷ 7 = ☐   70 × (53 − 38) = ☐

(42 × 8) − 138 = ☐   (675 − 451) ÷ 4 = ☐   160 − (160 ÷ 4) = ☐

(1250 − 250) ÷ 25 = ☐   105 × (43 − 37) = ☐   (52 × 40) − 96 = ☐

Time spent: _____ min _____ sec. Total: _____ out of 22

ANSWERS

Answers are given from top left, left to right, unless otherwise stated.

Test 1

Step 1:

33; 418; 514; 657; 446; 373; 188; 199; 1070; 801

Step 2:

1030; 100; 1090; 1300; 2400; 1000; 5650; 7460; 2390; 8309; 4781; 6077; 5356; 2400; 10 720

Step 3:

100; 1000; 100; 10; 1000; 100; 1000; 10

Test 2

Step 1:

First row: seven thousand three hundred and twenty-six; three thousand eight hundred and twenty-one; eight thousand two hundred and thirty; nine thousand five hundred and twenty-six
Second row: nine thousand six hundred and sixty-two; five thousand six hundred; three thousand seven hundred and two; eight thousand seven hundred and fourteen

Step 2:

7000 + 700 + 80 + 8;
1000 + 500 + 60 + 0;
2000 + 800 + 0 + 0;
3000 + 300 + 90 + 5;
9000 + 0 + 60 + 5;
4000 + 500 + 80 + 6;
5000 + 0 + 80 + 0;
3000 + 400 + 90 + 9;
7000 + 0 + 80 + 6;
8000 + 500 + 50 + 0;
6000 + 0 + 30 + 4

Step 3:

From top: 2100; 9850; 4626; 7003; 5030; 4106; 6820; 9400

Test 3

Step 1:

1. 3900, 3910; 5830, 5840; 4400, 4410
2. 3200, 3300; 4500, 4600; 5300, 5400
3. 9000, 10 000; 8000, 9000; 2000, 3000

Step 2:

First row: 4970; 3020; 2390; 8950; 9310
Second row: 3400; 4700; 7400; 6500; 5600
Third row: 5000; 3000; 3000; 4000; 6000

Step 3:

2300, 2350; 7485, 7385; 5753, 3753; 3256, 3356; 515, 715; 7100, 8100; 3812, 3813; 4350, 4460

Test 4

Step 1:

700; 750; 990; 500; 920; 710; 634; 679

Step 2:

715; 730; 936; 925; 688; 689; 878; 599; 998; 597; 786; 499; 682; 928; 909; 970; 802; 786; 964; 1053

Step 3:

1251; 1254; 1353; 996; 1225; 1292; 1715; 1934

Test 5

Step 1:

35; 26; 45; 36; 19; 39; 29; 39; 45; 19

Step 2:

222; 122; 313; 243; 431; 503; 422; 222; 338; 319; 227; 415; 458; 128; 279; 579; 141; 185

Step 3:

<; >; =; <; >; =; >; <; >

Test 6

Step 1:

210; 320; 950; 920; 790; 980; 290; 120; 240; 480

Step 2:

117; 859; 829; 210; 605; 505; 295; 610; 395; 140; 402; 896; 892; 672; 500; 176

Step 3:

210; 575; 220; 166; 670; 340; 545; 870; 575

Test 7

Step 1:

4600; 3400; 5900; 9100; 6900; 8400; 5600; 1700; 4400; 6500

Step 2:

First table, from top: 6700, 6692; 8600, 8594; 7800, 7793; 8000, 8032
Second table, from top: 2900, 2882; 4000, 4069; 6200, 6278; 3200, 3177

Step 3:

First table, from top: 4800, 4775; 9300, 9262; 2800, 2779
Second table, from top: 8300, 8271; 4800, 4877; 9300, 9262

Test 8

Step 1:

500; 4500; 6000; 1300; 3900; 9000; 9000; 8000; 9000

Step 2:

4900; 3900; 8400; 8900; 9700; 6800; 7700; 9900; 5900; 5800; 3600; 8800; 6800; 9400; 8700

Answers

Step 3:
3406; 5744; 6577; 5559; 5861; 5899; 8927; 4028

Test 9

Step 1:
200; 850; 200; 300; 500; 600; 100; 300; 300; 600

Step 2:
2100; 3500; 4100; 3300; 1500; 3000; 5400; 3200; 6000; 2400; 2600; 3300; 2200; 4400; 1800

Step 3:
3101; 5130; 3118; 2060; 4224; 2030; 2402; 3223

Test 10

Step 1:
80; 28; 52; 54; 91; 37; 790; 530; 260; 800

Step 2:
48; 28; 110; 120; 770; 75; 85; 162; 60; 680; 520; 48; 250; 1000; 309

Step 3:
100; 250; 160; 170; 22; 165; 1000; 21

Test 11

Step 1:
1; 5; 10; 50; 100; 500; 1000

Step 2:
8; 4; 12; 15; 43; 66; 9; 82; 37

Step 3:
>; =; <; >; <; =

Test 12

Step 1:
0; 0; 8; 79; 11; 10; 2; 6; 12; 60; 63; 48

Step 2:
12; 48; 18; 66; 30; 72; 42; 54; 4; 5; 3; 9; 8; 60; 1; 54; 11; 6; 36; 10

Step 3:
45; 60; 74; 59; 90; 11; 19; 75; 93

Test 13

Step 1:
32; 44; 8; 28; 9; 44; 30; 63; 54; 48; 71; 24

Step 2:
14; 28; 56; 35; 42; 14; 21; 28; 35; 84; 56; 63; 7; 70; 21; 0; 77; 42; 84; 49

Step 3:
3; 5; 7; 2; 8; 11; 12; 6; 9

Test 14

Step 1:
33; 18; 0; 81; 84; 72; 11; 22; 21; 0; 28; 12

Step 2:
18; 45; 81; 90; 54; 99; 63; 72; 36; 108; 0; 6; 2; 12; 9; 11; 7; 3; 5; 10

Step 3:
27; 9; 5; 63; 9; 90; 4; 9; 6

Test 15

Step 1:
55; 14; 69; 20; 50; 88; 17; 37; 100; 29; 27; 84

Step 2:
1. 8, 16; 9, 27; 10, 40; 6, 30; 9, 54; 6, 42
2. 56; 72; 49; 22; 60; 48

Step 3:
From top:
$10 \times 4 + 9 \times 4 = 40 + 36 = 76$
$10 \times 8 + 2 \times 8 = 80 + 16 = 96$
$10 \times 4 + 3 \times 4 = 40 + 12 = 52$
$10 \times 6 + 5 \times 6 = 60 + 30 = 90$
Other suitable answers are possible.

Test 16

Step 1:
100; 170; 325; 129; 1375; 170; 116; 369; 211; 520

Step 2:
225; 669; 780; 328; 296; 222; 1360; 156; 648; 322; 558; 234; 140; 496; 664

Step 3:
2100; 235; 91; 810; 221; 140; 47; 336

Test 17

Step 1:
104; 45; 103; 69; 312; 23; 32; 192

Step 2:
123; 234; 94; 144; 129; 159; 81; 19; 107; 140; 55; 54; 89; 69; 56

Step 3:
26; 108; 25; 34; 111; 93; 84; 81; 100

Test 18

Step 1:
900; 4200; 2000; 600; 4800; 5400; 2700; 1400; 3200; 2500

Step 2:
480; 940; 480; 570; 1000; 260; 960; 1100; 960; 910; 750; 990; 720; 510; 960

Answers

Step 3:

800; 500; 450; 520; 1440; 960; 680; 1280; 1050

Test 19

Step 1:

1. 24, 240   **2.** 72, 720   **3.** 26, 260

4. 160, 1600   **5.** 203, 2030

Step 2:

Alternative combinations are possible:

2. 20, 3, 980, 147, 1127

3. 35, 40, 35, 6, 1400, 210, 1610

4. 21, 10, 21, 8, 210, 168, 378

5. 29, 50, 29, 2, 1450, 58, 1508

6. 71, 30, 71, 4, 2130, 284, 2414

Step 3:

165; 275; 418; 891; 374; 154; 693; 1012

Test 20

Step 1:

1. 45, 450   **2.** 140, 1400   **3.** 128, 1280

4. 216, 2160   **5.** 108, 1080

Step 2:

960; 1120; 1140; 1290; 2160; 3150; 1200; 2200; 1110; 3200; 1890; 1560; 1120; 1890; 4080

Step 3:

2310; 2050; 870; 2240; 3350; 2070; 6880; 3840

Test 21

Step 1:

200; 40; 120; 50; 90; 60; 80; 80; 60; 30

Step 2:

20; 5; 10; 5; 60; 6; 8; 8; 7; 5; 9; 6; 9; 13; 4

Step 3:

9; 9; 7; 14; 8; 7; 6; 42

Test 22

Step 1:

212; 222; 192; 176; 427; 380; 234; 448

Step 2:

2. 522, 540   **3.** 752, 720   **4.** 1268, 1200

5. 1368, 1500   **6.** 3740, 3500

Other suitable estimates are possible.

Step 3:

168; 259; 424; 1704; 3468; 5312

Test 23

Step 1:

234; 420; 462; 552; 783; 544; 651; 498

Step 2:

2. 504, 540   **3.** 744, 720   **4.** 1482, 1500

5. 3402, 3500   **6.** 5824, 5600

Other suitable estimates are possible.

Step 3:

354; 584; 792; 1432; 4068; 7152

Test 24

Step 1:

1. 480, 8, 60   **2.** 300, 6, 50   **3.** 210, 15, 14

4. 960, 16, 60

Step 2:

96; 19; 40; 10; 5; 180; 2580; 125; 16; 15; 50; 24; 24; 18; 15

Step 3:

20; 60; 5; 625; 15; 40; 19; 224

Test 25

Step 1:

$\frac{1}{2}$; $\frac{1}{3}$; $\frac{1}{4}$; $\frac{1}{2}$; $\frac{1}{4}$

Step 2:

6; 5; 5; 5; 2; 4; 4; 4; 8

Step 3:

1. F   **2.** F   **3.** T   **4.** F

5. F   **6.** F   **7.** T   **8.** F

Test 26

Equivalent answers are possible.

Step 1:

$\frac{3+2}{7} = \frac{5}{7}$; $\frac{1+3}{5} = \frac{4}{5}$; $\frac{5+4}{10} = \frac{9}{10}$; $\frac{13+11}{25} = \frac{24}{25}$; $\frac{2+5}{8} = \frac{7}{8}$; $\frac{28+49}{123} = \frac{77}{123}$

Step 2:

$\frac{7}{9}$; $\frac{39}{58}$; $\frac{3}{4}$; $\frac{8}{11}$; $\frac{3}{4}$; $\frac{16}{19}$; $\frac{2}{3}$; $\frac{15}{17}$; $\frac{37}{47}$; $\frac{7}{8}$; $\frac{37}{39}$; $\frac{13}{20}$; $\frac{77}{108}$; $\frac{40}{209}$; $\frac{85}{107}$

Step 3:

$\frac{9}{10}$; $\frac{15}{17}$; $\frac{101}{137}$; $\frac{19}{29}$; $\frac{4}{18}$; $\frac{1}{24}$; $\frac{28}{225}$; $\frac{11}{108}$

Test 27

Equivalent answers are possible.

Step 1:

$\frac{9}{11}$; 1; $\frac{5}{6}$; $\frac{15}{19}$; $\frac{29}{31}$; $\frac{3}{20}$; $\frac{14}{41}$; 0; $\frac{1}{2}$

Step 2:

$\frac{15}{16}$; $\frac{11}{40}$; 1; $\frac{10}{13}$; $\frac{25}{26}$; 1; $\frac{17}{45}$; 1; $\frac{15}{37}$; $\frac{8}{37}$; $\frac{13}{30}$; $\frac{25}{74}$; $\frac{31}{47}$; $\frac{8}{29}$; $\frac{15}{43}$; $\frac{7}{18}$

Step 3:

$1\frac{1}{16}$; $\frac{73}{91}$; $\frac{19}{45}$; $\frac{61}{93}$; $\frac{23}{45}$; $\frac{40}{96}$; $\frac{7}{84}$; $\frac{23}{44}$

Answers

Test 28

Step 1:

Circled: 0.15; 7.23; 100.04; 0.3; 20.1; 0.04; 30.2

Step 2:

1, 50, one point five (zero); 4, 25, four point two five; 1, 20, one point two (zero); 2, 18, two point one eight; 49, 90, forty-nine point nine (zero)

Step 3:

From top: 2200; 7600; 4, 58; 85; 1600; 900

Test 29

Step 1:

Equivalent fractions are possible:

1. $\frac{7}{10}$, 0.7   **2.** $\frac{6}{10}$, 0.6

3. $\frac{4}{10}$, 0.4   **4.** $\frac{43}{100}$, 0.43

Step 2:

1. $\frac{3}{10}$; $\frac{6}{10}$; $\frac{9}{10}$; 0.1; 0.5; 0.9

2. $\frac{3}{100}$; $\frac{5}{100}$; $\frac{14}{100}$; 0.05; 0.08; 0.11; 0.18

Step 3:

0.29; 0.6; 0.718; 0.805; 0.1; 0.9; 0.143; 0.12

Test 30

Step 1:

>; <; >; <; <; >: >; <

Step 2:

>; >; >; >; =; >; <; >; =; <; =; <; >; >; <; >; >; =

Step 3:

>; <; <; <; <; =; =; <; >; =

Test 31

Step 1:

0.7; 0.03; 0.38; 3.92; 0.09; 0.21; 0.7; 0.05; 0.52; 5.42

Step 2:

<; <; <; <; >; =; <; <; <; =; >; <; >; <; <; =

Step 3:

1. 5.0, 5.05, 5.5, 5.55, 50

2. 0.33, 3, 3.03, 3.3, 33

Test 32

Step 1:

Right angles: 1, 3, 5, 7

Acute angles: 8, 9

Obtuse angles: 2, 4, 6, 10

Step 2:

Right angles: 3:00, 9:00

Acute angles: 10:00, 7:45, 5:30, 2:00

Obtuse angles: 9:30, 1:30, 11:30

Step 3:

∠ 2: obtuse; ∠ 3: right; ∠ 4: right; ∠ 5: right; ∠ 6: right; ∠ 7: obtuse

Test 33

Step 1:

1. line segments, triangle

2. four

3. line segments, pentagon

Step 2:

Triangles: 4

Quadrilaterals: 1, 3, 8, 9

Pentagons: 5, 7

Other: 2, 6

Step 3:

3, 3, triangle;

4, 4, quadrilateral;

4, 4, quadrilateral;

5, 5, pentagon

Test 34

Step 1:

1. acute, right-angled, obtuse

2. obtuse

3. acute

4. right-angled

5. scalene, isosceles, equilateral

Step 2:

Acute: 1, 2, 3, 4, 6, 10

Obtuse: 7, 9

Right-angled: 5, 8

Isosceles: 1, 3, 4, 5, 9

Equilateral: 1, 4

Step 3:

1. 2   **2.** isosceles

3. isosceles, equilateral   **4.** sides, angles

Test 35

Step 1:

Shapes 1, 2, 3 and 5 ticked and lines of symmetry drawn

1.

2.

3.

5.

Answers

Step 2:

Shapes 2, 3, 4 and 6 ticked and lines of symmetry drawn

2.

3.

4.

6.

Step 3:

1. 2; 4   **2.** 3   **3.** 1   **4.** infinite

Test 36

Step 1:

1. length, width; 15

2. side length, side length; 16

Step 2:

1. 35; 300; 1750; 2940; 2700; 800

2. 225; 6400; 121; 400

Step 3:

1. 35 000; 3200; 8000

2. 250 000; 490 000; 160 000

Test 37

Step 1:

1. m   **2.** km   **3.** m   **4.** km

5. m   **6.** km   **7.** km   **8.** m

Step 2:

1000; 4000; 30; 9000; 5; 18; 7; 80 000; 38; 19 000; 8750; 7007; 99; 16 040; 3450

Step 3:

5020; 9067; 4300; 7800; 7570; 7920; 29 400; 2

Test 38

Step 1:

1. perimeter

2. 2 × length + 2 × width (in either order)

3. 4 × side length

4. 25; 26; 26; 28

Step 2:

First table: 56 cm, 90 cm, 150 cm, 184 cm, 360 cm

Second table: 136 cm, 200 cm, 288 cm, 352 cm, 172 cm

Step 3:

From top:

1. 16; 26; 12; 32

2. 19; 23; 16; 12

Test 39

Step 1:

60; 60; 24; 7; 12; 3600; 100; 120

Step 2:

240; 210; 1.25; 36; 24; 16; 6; 80; 950; 0.70; 14; 6.08

Step 3:

11, 16; 5, 90; 36, 2160; 0.75, 2700; 28, 672; 30

Test 40

Step 1:

15; 232; 240; 32; 693; 106; 761; 1480

Step 2:

(930 − 690) ÷ 5 = 48;

523 − (609 ÷ 3) = 320;

(36 × 23) + 123 = 951;

1008 − 41 × (15 + 8) = 65;

(979 −115) ÷ (3 + 13) = 54

Step 3:

87; 51; 1050; 198; 56; 120; 40; 630; 1984